Introduction

This book provides essential information and guidance for mouse owners to ensure the well-being and happiness of their pet mice. The guide covers a wide range of topics related to mouse care, including their introduction, daily care, diet and nutrition, handling, health, breeding, and general mouse behavior.

The guide starts with an introduction to mice as pets and provides an overview of the different types of mice, including fancy mice. It discusses the pros and cons of having mice as pets and explores the various body types, coat varieties, and self-colors found in fancy mice.

For those considering getting a pet mouse, the guide offers tips on choosing a healthy mouse, whether to have one or two mice, and how to transport and transition them to their new home. It emphasizes the importance of daily mouse care, including selecting the right habitat, proper diet and nutrition, providing toys for stimulation, and addressing hygiene needs such as bathing.

The guide also covers important aspects of mouse health and breeding. It provides guidance on observing mice for signs of illness, common illnesses and problems, and the issue of zoonotic diseases. For those interested in breeding mice, it discusses fertility in females, determining mating, pregnancy and birth, and an introduction to understanding genetics.

Additional topics addressed in the guide include rodent shows, taming wild mice, the suitability of feeder mice as pets, the benefits of fancy mice as pets, and frequently asked questions such as the number of mice to get, gender distinctions, introducing new mice, and the level of care required for pet mice.

Throughout the guide, care summaries provide quick reference points for important aspects of mouse care, such as picking a pet, mouse hygiene, signs of ill health, habitat recommendations, nutrition, and maintenance.

The guide concludes with a section on the representation of mice in popular culture, highlighting their presence and significance in various media.

Overall, this book serves as a comprehensive resource for mouse owners, offering practical advice, tips, and information to ensure the well-being and happiness of pet mice.

Contents

Chapter 1 - Introduction to Mice

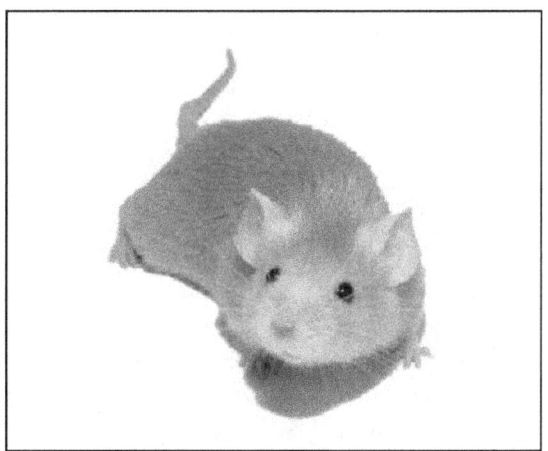

Perhaps you've heard all the usual phrases, "Quiet as a church mouse," or "timid as a mouse." Maybe you've known someone who was so tied to home and hearth they were described as a "house mouse." And then, of course, there's that old worn-out saw about the best laid plans of mice and men.

Although often portrayed as pests, the destroyers of grains and the raider of larders, mice are also well beloved as tiny companion animals. Packed with personality and "packaged" in a variety of looks, you may be surprised to discover just how fascinating the world of the humble mouse can be.

What are Rodents?

Rodents are mammals with distinctive upper and lower incisors that are well adapted for gnawing and that continue to grow throughout the animal's life.

They use their teeth to break down their food, chew through dense materials to make their homes, and as a defense against predators. Male mice can be highly territorial and can do serious damage to one another with those teeth!

Approximately 40% of all the mammals on earth are rodents. They can be found on every continent, but not in Antarctica. Some of the most commonly recognizable are squirrels, rats, and mice, but porcupines and beavers are rodents as well.

What are Mice?

Mice are among the smallest of the rodents, and arguably some of the cutest, so much so that they have found their way into popular culture as literary and motion picture characters.

They have tiny, pointed little snouts flanked by luxuriant whiskers, and long tails that are either naked or almost hairless. Of the many species of mice, the best known is the common house mouse, *Mus musculus.*

Described as having an agouti coat, meaning each individual hair is ticked with bands of color, the common house mouse appears to be brown, usually with lighter shading on the muzzle and perhaps on the feet.

All mice have whiskers or "vibrissae" located on their upper lips and snouts, with one whisker under each eye. There are shorter whiskers on the chin.

All of these specialized hairs are navigational aids, allowing the mouse to sense movements in the air and around objects nearby as compensation for their less than remarkable eyesight. They are also aided by an unusually keen sense of smell.

Mice have four toes on each front foot and five toes on each back one. They use their front feet to hold food, so that the tiny paws look almost like hands. The hind feet give the mouse stability, allowing him to stand upright when something catches his interest.

The long tail serves as a counterbalance, making possible a scampering mouse's acrobatic feats of climbing and leaping. Mice can traverse even the wobbliest surface with utter confidence, and even use their tails as a sort of fifth leg when climbing.

The Mouse Fancy

Mice are believed to have originated in Asia and it was there that the Chinese first kept them as pets as early as 1100 BC. Archaeologists have also found drawings of colored mice from Egyptian tombs of roughly the same period.

The Egyptians believed mice to be in possession of supernatural powers, while the Greeks used them as aids in medicine and for divination and prophesy.

These first Chinese mice were likely captured from granaries and households and may have been spared a worse fate because they were in some way unusual and thus interesting. There are many references to white mice having been captured from wild circumstances.

As a species, mice adapt quickly to domestication, becoming calm and social. Consequently, selective breeding in captivity led to the development of many different coat, size, and texture variations.

In Japan in the 18th century the booklet, "The Breeding of Curious Varieties of the Mouse" led to the selective breeding of varieties including the Albino, Black-Eyed White, Chocolate, Lilac, and Champagne. A mouse was believed to be a messenger portending the arrival of wealth.

As unusual mice found their way to Europe, they attracted enthusiastic attention. The "mouse fancy" became popular in England in the late 19th century, and the terms "fancy mice" and "hobby mice" began to be used to distinguish the more "civilized" variations from their plainer, country cousins.

The recognized "father" of the hobby, Walter Maxey, assisted in the formation of the National Mouse Club in 1895 and the subsequent development of standards by which mice were judged in organized shows, starting that same year.

At the same time that mice were being kept as pets, the animals came to the attention of scientists for use in research, including the work of Gregor Mendel in genetics. These experiments created even more unusual specimens that ultimately found their way back into the pet trade.

The spread of the mouse fancy to the United States is not so well documented, and the first pets were likely either wild mice or laboratory animals.

The American Mouse Club was formed in the 1950s, but languished from disinterest. In 1978, however, the American Fancy Rat and Mouse Association took its place and remains an active organization today with an enthusiastic membership.

Types of Mice

In his most basic, brown format the mouse is said to have an "agouti" or ticked coat. From there, however, breeders have cultivated almost innumerable shades that are separated into the following color categories:

- selfs, all one color

- tans, one solid color on top with a tan belly

- patterns, even or broken

In terms of their origin, mice are broken into four groups: feeder, fancy, laboratory, and wild.

Feeder Mice

In the United States, feeder mice are the most readily available, but the practice of selling live mice for reptile food is illegal in the United Kingdom.

People who do raise mice to be marketed as reptile food breed the animals randomly with little if any concern about inbreeding and genetic defects. It is also quite common for feeder mice to be kept in filthy, deplorable conditions with sub-standard nutrition.

Feeder mice are, however, quite inexpensive, selling for $0.30-$0.50 / £0.18-£0.30 each. In most cases feeder mice are only available in large lots, but it is often possible to talk a sympathetic pet store manager into selling you one or two.

If you are lucky enough to get healthy specimens from this kind of purchase, they can, with love and attention, be good pets, but most still die within the first year of life.

Take this into consideration, especially if you are purchasing the mouse as a pet for a child. The loss of a pet so quickly after it has joined the household can be devastating for your son or daughter.

Fancy Mice

The world of fancy mice is too extensive for a brief explanation, and for this reason the following chapter will be dedicated exclusively to these elites of the mouse world. You'll be shocked to find out just how princely some of these tiny creatures can be!

Laboratory Mice

If you have access to someone who works in a lab and can get ahold of either the BALB/c or C57BL/6 strains of lab mice, they make excellent pets.

The BALB/c is an albino strain of the common house mouse that is now more than 200 generations old. It was first developed in New York in 1920, and has been distributed globally as a laboratory animal.

The C57BL/6 is widely used for its suitability in testing with human disease strains. They reproduce easily and prolifically and are extremely robust little animals.

Wild Mice

In most cases, we encounter wild mice at the very times and places where we don't want to see them — like the pantry. These creatures do not tend to make good pets.

By instinct they are wary animals that never seem to be able to calm down no matter how kindly they are treated. In addition, wild mice can carry disease transferrable to humans. (See the chapter on health for more information on zoonotic diseases.)

This is why so many people harbor the stereotype that mice are filthy. The farthest thing could be true of companion mice. They are actually rather fussy about their homes, which they tend to arrange and rearrange constantly to keep things in good order.

Mice as Pets

As pets, mice are gentle and loving, with a curious streak that is both eerily intelligent and thoroughly beguiling. They are quiet and inexpensive to both feed and house. They enjoy interaction with their humans and are friendly to the point of being gregarious.

The major drawback to welcoming a mouse into your life is their brief lifespan, just 740-1000 days. The little creatures only weigh 1-2 oz (28.3-56.7 g) and are quite fragile.

Because they are nocturnal, it's not unusual for mice to stage nighttime escapes if given less than half a chance although this tendency lessens over time.

Don't make the mistake of keeping your pets in the bedroom with you. You'll quickly discover they have two speeds, stop and go — and when they're "going," it's a mile a minute. I have often wished that I had a fraction of the energy my mice exhibit on a perfectly normal day!

Heaven help the mouse owner sharing sleeping quarters with a pet running in a squeaky exercise wheel or a rattling water bottle. They will keep you up all night with their busy nocturnal habits.

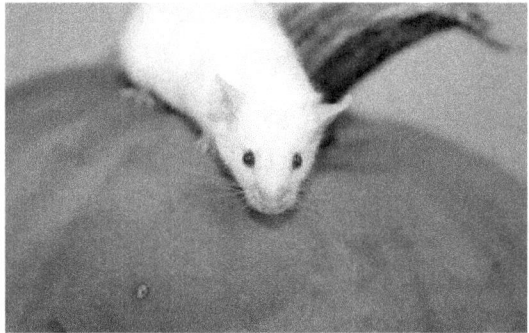

Regardless of where you put their habitat, however, you'll find that odor is not a problem with pet mice. They are so fastidious they will pick one corner of their cage to urinate and defecate making spot cleaning extremely simple for their keepers.

More on Life Span

I think it's important to discuss life span at somewhat greater length, especially if you are considering getting mice as pets for your children. It is quite rare for a mouse to live more than three years and many die within the first year of life.

How you care for your pet mice will greatly influence their longevity. If you restrict their calorie intake, mainly by not overfeeding with treats, they will live longer. Chubby mice die much faster.

For this reason, it's important for parents to be involved in the husbandry of mice kept by children. Explain proper dietary requirements and discourage your son or daughter from offering treats to their mice.

Regardless, you must prepare your children for the inevitable. Their mice will die quickly. For this reason, it's actually best to keep multiple mice. This helps to lessen the upset when a member of the group is lost.

Mice and Allergies

Mice are not hypoallergenic, and in fact are responsible for more allergic reactions in humans than either dogs or cats. If you are looking for a pet for a child with severe allergies, you would be much better off getting a turtle or some sort of lizard. (Although parents should also understand that these types of pets can carry the salmonella bacteria.)

It's also important to understand that allergies are highly species specific. The reaction is actually to proteins present in the salivary and sebaceous glands of the animal, which are transferred in the dander created from self-grooming or distributed in oils applied for purposes of marking territory.

A child who is allergic to cats may be fine around a dog, bird, or mouse. Have your child thoroughly tested before acquiring a pet of any type. The adverse reaction is not the animal's fault and a lot of heartbreak can be avoided by getting all the facts about the allergy well in advance of any planned adoption.

Mice and Children

In evaluating the suitability of mice as pets for your child, consider your child's level of maturity and degree of responsibility. Mice are delicate. They can be easily injured. They must be fed appropriately, and their habitats must be kept clean.

In almost all cases I do not recommend giving a child under the age of 9 sole responsibilities for caring for pet mice. Even at that age, it is highly recommended that parents still exercise an observational if not supervisory role in the care of the little creatures.

Children should be well educated in the necessity of being kind and gentle with all animals, not just in the matter of handling, but also vocally.

Mice do not see well, so their hearing is extremely acute, and they can become dangerously terrified and stressed by loud noises.

Make sure that your children always wash their hands before and after handling their pets.

Mice and Other Pets

I am always a little taken aback when someone asks me if a mouse will get along with a cat or a dog. It is almost all I can do not to say, "Yes, until your pet eats the mouse." So there. I've said it.

You have to be realistic. Mice are prey. Even your very well-mannered parrot may decide to make a snack of your pet mouse. This is not a situation where mixing is even mildly an option.

Mice must be kept well away from all other domestic animals and their habitat must be secured against invasion from other pets. If not, the consequences are assured and unpleasant.

Pros and Cons of Mice as Pets

All of the following are more or less "cons" for having a mouse as a pet, although most are not what anyone would regard as a "deal breaker."

- Mice have short lives.

- Mice are trainable, but they are not interactive in the same way a dog or a cat would be.

- Mice must be kept in a habitat, and not allowed to roam loose in the house.

- Mice are nocturnal, so they are primarily active while you are asleep.

- Mice are very fragile and can be injured easily.

- Mice can trigger severe allergic reactions in some people.

The "pros" of pet mice are fairly clear:

- They are inexpensive pets.

- They require minimal housing like an aquarium, wire cage, or specially designed maze-like environments.

- They are cheap to feed and will thrive on readily available pellet foods.

- They make no loud noises, although they are primarily active at night and should not be kept in bedrooms.

- They are fastidiously clean so that their habitats will not emit foul odors if well kept.

- They are cute and intelligent and come in many colors and patterns.

- If you are interested in cultivating a more active hobby along with pet ownership, fancy mice can be shown competitively in exhibition.

Almost any of these points could be argued as either a reason to get a mouse or a reason not to. "Pros" and "cons" of pet acquisition are truly a matter of personal preference. For some people mice are absolutely perfect pets because they are smart, interactive, and easy.

Chapter 2 - The World of Fancy Mice

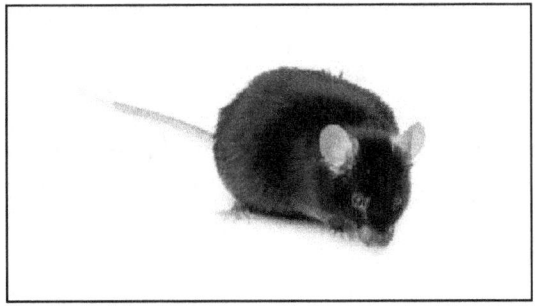

The world of fancy mice is indeed so "fancy" as to almost be overwhelming. If you have not done so, seize the opportunity to attend a rat and mouse show to truly appreciate the variety of looks and body types in animals commonly perceived to be "plain." As you're about to find out, there's nothing plain about fancy mice.

Body Types

Mice are separated into three distinct body types, English, tailless, and gremlin.

English

Oddly enough, English isn't really a body type at all, but is simply lumped into this designation for convenience sake. In truth, an English mouse can have any body shape, but it must have English blood.

Tailless

As the name clearly implies, a tailless mouse has no tail. The preference, for show purposes, is that there be no tail at all, but if a shortened tail is present, it must be straight and without kinks or abnormalities.

Gremlin

In gremlin mice, one ear is in the normal location, while the other is placed on the side of the head. The look is both distinct and whimsical.

Coat Varieties

In fancy or show mice, there are 12 possible coat varieties: standard, satin, angora, long hair, rex, caracul, texel, frizzy, fuzzy, hairless, rhino hairless, and rosette.

Standard

Mice with standard coats have a very high shine. Their hair is short, but soft, smooth, thick, and glossy.

Satin

In satins, all parts of the hair are thinner. This variation can be paired with other coat types so that you get combinations like a Satin Rex or a Satin Angora. When satins are babies, their fur comes in with a higher than normal sheen. On darker mice, this may be difficult to detect unless you look at their bellies.

Angora

In angoras, all coats are longer than normal because the outer or guard hairs are longer and seem wool-like with a pattern that zigzags.

This tendency for a longer coat becomes apparent at about 18 days of age although the coat may shorten some before the animal reaches adulthood.

As with the satins, all combinations are possible so that you will see designations like Angora Fuzzy or Angora Rex.

Long Hair

In long hair mice the outer guard hairs are exceptionally long, but unlike angoras, there is no zigzag pattern present. This tendency for long hair becomes visible at 4-5 weeks of age.

The head, feet, and tail will look more like a standard coat, and all combinations are possible, for instance Long Hair Rex or Long Hair Satin.

Rex

Rex mice have whiskers that are crimped or curled at birth. As their fur develops, it lays in waves that straighten only slightly as the animal ages. Typically, the coat passes through three stages, curly, standing on end, and wavy.

In combination, the coat types may be described as Angora Rex, Satin Rex, and so on.

Caracul

The caracul coat is similar to that of the rex. The mouse's whiskers will come in at birth curled or will start to curl after a few days.

The fur has a noticeable wave, but straightens out a great deal by about four weeks. As adults, the coat is plush with very little wave or curl visible.

Texel

The coat of a texel mouse has a tight curl across the entire body with curly whiskers. The guard hairs are especially thick, curled, and dispersed throughout the coat and remain so throughout adulthood.

Frizzy

Frizzy mice are not so much curled as crimped. Their hair is coarser. The effect of the frizziness is much easier to feel than to see.

Fuzzy

Fuzzy mice can have very little hair at all, to very thick and curly coats. All, however, have crimped or curled whiskers.

Hairless

Hairless mice are not only hairless, but they have no whiskers. Their skin should be bright and almost translucent with no scars or marks.

Rhino Hairless

The rhino hairless should be identical to the hairless, but with large ears and deeply wrinkled skin like that seen on Shar Pei dogs.

Rosette

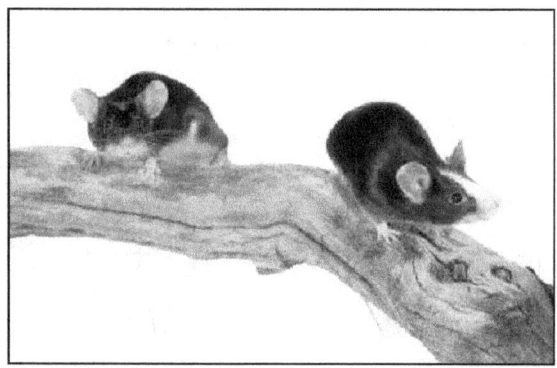

Rosette mice have a whorl of fur on each hip that appears to be a spiral pattern in opposition to the standard direction of the hair.

Self Colors

Mice that are all one color are said to be "self" colors. There are 19 variations including:

- non-agouti (black)

- extreme non-agouti (extreme black)

- chocolate

- mock chocolate

- light mock chocolate
- champagne
- coffee
- beige
- lilac
- blue
- silver
- dove
- lavender
- recessive yellow
- yellow
- cream
- albino & PEW (pink eye white)
- BEW (brown eye white)
- REW (ruby eye white.)

Non-Agouti (Black)

Non-agouti or black mice are jet black with dark eyes, but with yellow at the base of the tail, around the ears, and at the genitals and nipples. Most of the yellow hairs fall out when the mouse reaches adulthood.

Extreme Non-Agouti (Extreme Black)

Extreme non-agouti mice are even blacker in appearance, and do not display the yellow coloration on some areas of the body as is seen with the non-agouti type. They also have dark eyes.

Chocolate

Chocolate colored mice have coats that are a rich, dark brown with matching brown eyes. They do look as if they were made of the finest dark chocolate.

Mock Chocolate

Mock chocolate mice are still dark, but they lack the deep coat tones that are seen in the more intense chocolate variety. Mock chocolate mice still have dark brown eyes, however.

Light Mock Chocolate

A light mock chocolate mouse is a shade lighter than a mock chocolate. Often this is a difficult distinction to make without a point of comparison to the other designated chocolate shades.

If you were to place chocolate, mock chocolate, and light mock chocolate mice side by side, they would look like gradually lightening paint chips. Each coloration has dark eyes, however.

Champagne

The champagne coat is an even lighter brown with just a hint of pink in the undertone. Unlike the darker brown coats, champagne mice have pink eyes.

Coffee

The coffee coloration is a softer brown than champagne, and these mice have dark eyes. They look something like the color of a cup of coffee with a bit of cream added.

Beige

Mice that are beige appear to be a shade between an off white and an actual tan. In general, this is a warmer shade, while tan is somewhat flatter. Their eyes are dark.

Lilac

The lilac coat appears to be almost blue, but with a vaguely pink tinting that can create a slight suggestion of light purple, hence the name. The eyes in this coloration are pink.

Blue

A blue mouse's coat is much deeper than that seen in a lilac, to the point of being a deep slate with less of the suggestion of purple shading. Mice that are termed to be "blue" will have dark eyes.

Silver

Mice that are silver have a light grayish appearance almost like the tonality of chrome. Their eyes are always pink.

Dove

The dove coat is almost an even mix of blue and chocolate. The eyes in this coloration will be dark.

Lavender

Lavender mice appear to be a mix between the silver and champagne coats. Their eyes are pink.

Recessive Yellow

Mice who are said to be recessive yellow can take on any shade from vibrant red through light blond and into dark sable. In youth they may have an uneven, sooty color with some dark ticking, but this goes away as they get older.

Their eyes may be any color, but those individuals that are red will almost always have pink eyes, while the blonds and sables exhibit a dark eye color.

Yellow

Recessive yellow, which is often called Lethal Yellow, exhibits across the same wide range of colors, but they are distinct in their genetic difference and may exhibit white spotting.

Cream

Mice with a cream coat appear to be a light, slightly yellowish shade with dark eyes.

Albino & PEW (Pink Eye White)

Albino & Pink Eye White (PEW) mice appear to be pure white because their coats have no pigmentation. Their eyes are always pink.

BEW (Brown Eye White)

Brown Eye White (BEW) have dark eyes and a pure white coat. They do have pigmentation in their coats, and are not classed as albinos.

REW (Ruby Eye White)

Ruby Eye White (REW) are also pure white (not albinos), but have very dark ruby eyes.

(It may be necessary to examine them in good lighting to distinguish a REW from a BEW.)

AOC (Any Other Color)

Because so many color combinations are possible in mice, these are the shades that are set apart from the self colors for show purposes.

Agouti

The agouti coat is an overlay of golden tan with a slate blue under color with the darkest hair on the back, lightening over the sides and belly. The eyes are dark.

Cinnamon

The cinnamon is very like the agouti, but the darkest color is more chocolate than slate, with the same lightening over the sides and down to the stomach. The eyes are also dark.

Blue Agouti

Another variation on the agouti coloration, these dark eyed mice show a color gradation from dark blue at the base of the hair shaft, through blue in the middle to white at the tip. This basic coloration can present with a range of tones, however.

Argenté

In the argenté, the top coat is yellow with a lilac under color. As babies, it's difficult to distinguish an argenté from other yellows until the lilac undercoat comes in.

Silver Argenté

In the silver argenté, the hairs are banded from an undercoat of slate blue to silver in the middle and white at the tip. These colors appear across a range of tones and it is sometimes necessary to breed selectively to eliminate browns.

Chinchilla

On a chinchilla mouse each individual hair should be blue at the base, gray in the middle, and black at the tip. The belly is fox and the eyes are dark.

Silvered

In silvered mice the hair may be completely white with no pigmentation, fully colored including black, colored with white tips, or banded.

Silvered mice can come in any color, and males will often be more silvered than females.

AOCP (Any Other Color Pattern)

The following mice not only display colors outside of the self range, but are further characterized by distinct patterning.

Brindle

Brindled mice are tiger striped from head to tail with less distinct stripes on their stomachs. Brindles are often used in labs conducting cancer research due to their propensity to grow tumors.

(This does not mean they are predisposed to cancer, only that they will readily develop tumors when cancer cells are introduced into their bodies.)

Roan

The roan coat mixes white with any other color in an even distribution throughout the body with greater amounts of white appearing on the belly. The eyes should be a match for the color of the body.

Merle

The merle coat has a marble-like pattern comprised of solid patches on a lighter roan base color. The eyes should be a match for the base color.

Himalayan

Himalayan mice are white with pink eyes and distinct dark points at the feet, ears, nose, and tail. For this reason, they are often confused with the color point beige.

Color Point Beige

The color point beige coat is much like that of the Himalayan, but the points at the feet, ears, nose and tail are darker and the eyes are darker.

Additionally, the body of a color point beige is creamier rather than pure white.

Siamese

The Siamese coat is another pointed variation, but the areas of shading are larger and the body creamier than the color point. Also, these mice have ruby eyes.

Given the name, it is not surprising that these mice look a great deal like Siamese cats in their coloration and markings.

Reverse Siamese

A reverse Siamese is a mouse of any color that has white points at the feet, ears, nose, and tail. They are often coffee shaded with white pointing.

Burmese

A Burmese mouse can be of almost any color, but it must have dark points on the feet, ears, nose, and tail that are the same color as the body, just darker.

Sable

A mouse is said to have a sable coat when the hair is dark on the back and fades to a reddish tan on the underside.

Splashed

Splashed mice are colored, with darker splashes on the body. They can come in any color and their eyes should match their predominant color. They should not be confused with mice that are variegated or brindled.

Tan and Fox

Tan and fox mice have distinct lines of tan with lighter bellies. These coat regions are highly delineated, almost as if they had been drawn in place.

Tan

A tan mouse's underside will be a dark golden red. They show a definitive line at the point in which their top color touches the bottom color.

This should run straight down the jaw, along the chest, and the sides. The shades of tan will vary and the eyes should match accordingly.

Fox

The fox coloration is very similar to the tan but the belly is pure white. Otherwise, the fox can be seen in all colors and patterns.

Marked

In addition to color and patterning, other mice display specific markings that are described by words like "banded" or "belted" that create a clear visual. Some markings are harder to understand, however, without extensive experience.

Belted

Belted and banded mice are very similar, both with wide markings around their mid-sections. The white band on a belted mouse begins on the back and is thicker there.

It then continues down the sides and around the belly, where as a banded mouse will have a thinner belly stripe. If well bred, however, a belted mouse can have an almost perfectly even band of white encircling the body.

Banded

A banded mouse is solid in color with a white band around the midsection that is usually thinner on the belly, but can also be broken without connecting at the spine.

Often the band is so wide it covers half the mouse, and the marking can also be a double band.

Piebald

Mice that are piebald have markings that are like those seen on black and white cows.

Dutch

Dutch mice have a patch shaped like an oval that begins at the front of the eye and continues to the back of the ear on either side of the face. The mark does not, however, touch the whiskers.

A white stripe may also be apparent between the darker face markings that begin at the nose and go around the ears. Dutch mice can be any color and their spots should be very clean cut.

Broken Merle

The broken merle mouse is a combination of roan and merle patches and white spots throughout the coat.

The merle and roan patches should each cover half of the body. This distinct pattern can be present with any color, and the eyes should match accordingly.

Broken Tan

A broken tan mouse is basically a tan mouse with spots. The body can be any color, so long as there is tan on the belly and white spotting is present. The eyes of a broken tan mouse should always match the body color.

Variegated

Variegated mice are also technically spotted, but the edges of the spots present are extremely jagged. The marks appear throughout their coats, which can be of any color with eyes to match.

They should not be confused with mice that are said to be splashed, which is a color on color pattern. Variegation is a pattern of white spots on a colored mouse.

Rump White and Colored Rump

As the name implies, these mice have a white rump with the rest of the body showing a distinct color. The Rump White has a line between the two areas that should be well defined and even around the body.

Rump Black and Colored Rump

The Rump Black is identical to the rump white with the exception that the posterior portion of the mouse is black with a distinct line around the body before the anterior color begins.

Tri - Color

The most simple explanation for a tri-color mouse is that it has three "colorations" but those can be a mix of splashes, spots, variegations, bandings, beltings, and colors. In most cases, however, "tri" mice are black, brown, and white.

Chapter 3 - Buying Mice as Pets

The healthiest mice with the longest life spans will come from mouse breeders, not from pet stores. Sadly, however, that is where most families go to get pets for their children.

While rescuing a feeder mouse from its intended fate is commendable, I recommend you find a dedicated individual breeder. Mice do not live long in the first place, but feeder mice are rarely healthy.

Your two best sources of contact information for breeders are the American Fancy Rat and Mouse Association at http://www.afrma.org in the United States and The National Mouse Club in the United Kingdom at http://www.thenationalmouseclub.co.uk.

Both organizations maintain breeder directories and can help you to find someone in your area from whom to purchase a pet.

You will pay from $10-$30 / £6-£18 or more for a fancy mouse, but you will get a genetically superior pet that is healthier and more socialized than a feeder mouse from a big box pet store.

Picking a Healthy Mouse

Since mice are nocturnal, try to visit a mousery in the late afternoon when the animals are beginning to show signs of greater activity. This will help you to observe the mice moving around, so you will stand a better chance of picking a healthy, well-socialized individual.

The following tips will help you to select a pet mouse that is healthy, friendly, and social.

- Do not pick the largest or smallest mouse in the bunch.

Even if the mouse is to be a child's pet, don't let your son or daughter make the selection. You should pick the mouse based on a clear evaluation of its health and personality. The smallest mouse may be cute, but the runt of the litter is more likely to have genetic weaknesses.

- Look for the mouse that approaches you first to sniff your fingers.

Mice should be active and interested in the world around them. The ones sitting off to the side that seem calm are actually not normal!

Mice that exhibit no fear and come right over to check you out will make good pets. These animals will be most receptive to socialization and daily handling.

- Do not select mice that have any discharge or discoloration around the eyes or nose.

Healthy mice have clean, shiny coats. Their eyes are wide open and bright. They show no evidence of nasal drainage, and no staining around or under the tail.

- Hold the mouse close enough to listen to it breathe.

Respiratory infections are a leading cause of early death in companion mice. When you listen to a mouse breathing, there should be no wheezing or sneezing evident.

- Pay attention to the condition of the mousery.

Ask to be given a tour of the whole "mousery" to get a sense of the conditions in which the mice are housed and bred. If the breeder refuses, try to determine if this is because they are concerned about spreading germs around their pets.

If the owner does show you around, pay close attention to the condition of the habitats. There should be no overpowering odors especially that of ammonia, and the mice should not be housed in overcrowded conditions.

Each cage should have clean bowls of food and water, and fresh, dry bedding. There should be adequate gnawing toys, and those that provide exercise and intellectual stimulation.

Consider the physical appearance of the mice, and try to observe the stools they have deposited. The pellets should be solid and well formed with no sign of diarrhea.

Mice are extremely clean animals and should be kept in well-maintained habitats. The bedding should be fresh, and there should be no odor.

Recommended Purchase Age

It's best not to buy a mouse that is under five weeks of age. After that time, they are quite capable of living independently and doing well without their mothers.

Do not, however, buy a mouse of more than 9 weeks of age unless you can verify that the animal has been handled and socialized.

If a mouse does not receive adequate human interaction before 9 weeks, they will tend to be antisocial and nervous for the rest of their lives. A mouse reaches full maturity at 12-16 weeks of age.

One Mouse of Two? Male or Female?

Two mice will always be happier than a single mouse living alone. These little creatures do get bored and lonely, leading them to act out their feelings in behaviors that are understandable if you realize what's going on.

A depressed mouse will be aggressive, and when he's not being ill-tempered, he'll sleep a lot. Think about it. The little fellow is unhappy and he's trying to tell you so!

I always recommend getting a pair of mice since having two does not significantly increase the level of care required. Your best option is to buy two females of the same size. Females are gentler, and their urine has a much less intense odor. Additionally, females tend to live a little longer.

If you house two males together, they will many times display territorial aggression and wind up hurting one another. The only time I would recommend getting two males is if they are littermates that have never been separated.

Never introduce a second adult male into an existing habitat with another adult male unless you want to see what two mice look like in a full out fight. It is not a pretty sight, and someone will get hurt.

Also, as a word of caution, don't try to reach into the habitat and attempt to separate two fighting mice. Use a fairly thick piece of cardboard as a barrier to place between the two mice and wear a glove to pick up one individual and remove it from the habitat.

Mice are not normally aggressive, but when they are angry, they will bite the hand that feeds them — and those little teeth are very, very sharp.

Beyond these housing considerations, there is no difference in the genders when it comes to their personality as pets, or to their health.

The male/female question is purely one of peaceful housing, and ensuring that you don't get any unplanned litters of pups.

Determining Gender

If you do get your mouse at a pet store, don't think for one minute that the employee will correctly identify gender. You can trust a breeder, and over time, you will likely learn how to make the distinction yourself.

The easiest way to explain the difference in the genital area is that female mice have two openings that resemble a "dash" near the body and an "O" near the tail. Both openings in males are circular.

After 14 weeks of age, however, the difference is quite clear since the scrotum and testicles of males are visible by that time. Unfortunately, that's a bit later than recommended to purchase a mouse unless it's been well socialized.

It's generally unwise to purchase a female mouse of more than 6 weeks of age that has been kept in a mixed population. Chances are very good she's already pregnant. This is especially true of pet store mice.

Transporting Your Mouse

Mice are very easy to transport. In most cases you'll bring your little pet home in a cardboard box. You can, however, purchase a small "pet keeper" with a cover and handle to bring your mice home. These plastic units are very inexpensive, usually selling for under $10-$20 / £6-£12.

Just in general I think it's a good idea to have one of these carriers on hand as a secondary habitat for your mice during times when you're cleaning the cage or otherwise need to move your pets.

It's certainly not out of the question to travel with your mice, but like any other pet, they must not be left in a vehicle that is not temperature controlled.

I wouldn't recommend taking your mice on a driving trip, but if you were planning on being in a vacation rental for a couple of weeks, it's certainly feasible to take the mice along and simply let them live in their travel box for that time with all the appropriate items for feeding, watering, and entertainment provided.

Your new pet's primary "house" should be all ready for moving in when you take your mouse home for the first time. The following chapter discusses setting up a habitat as well as the requirements for daily care.

The Transition to a New Home

Even though mice are highly social, they can react to the trip to a new home with some degree of stress, which is unhealthy for the sensitive little creatures.

I recommend letting mice have about 24 hours to get settled in before you try to handle your new pets all that much. You certainly do not want to try to handle them in the car on the way home. The mice will be nervous, and once they get away from you, catching them again is all but impossible.

When you arrive home, place the travel container inside the main habitat and let the little animals come out on their own. Make sure they have food, water, and a nest box. Everything about their new home should already be in place before you even buy your mice.

Be careful not to "loom" over your new pets. Watch how the mice are getting settled in from a distance and talk softly to them to let them get used to the sound of your voice.

Mice are interested in what's going on around them, so you will want to put their habitat in a part of the house that is active and not isolated.

However, for the first day or so, limit loud noises like a blaring TV set or stereo. Let the mice get used to the sounds of their new home. By the second day you can begin putting your hand in the habitat and letting the mice smell your fingers.

Rather than picking up mice, I prefer to let them step into my hand on their own, which most are very willing to do. Just place your hand palm up on the floor of the habitat, perhaps with a little treat to entice the mouse to step up.

Mice learn very quickly. After you've done this a time or two, your mouse will get "on board" whether you offer him a treat or not. Then you can handle your mouse without startling it in any way.

Most mice are very amenable to being held, just remember how tiny and fragile they are. Be gentle!

By nature, mice are fairly territorial, so you want to give them an opportunity to establish their habitat as "home base." This gives them a sense of security, so that after a few days, they really aren't interested in getting out, especially if they have other mice to keep them company.

Chapter 4 - Daily Mouse Care

One of the reasons mice have grown in popularity as companion animals is their low maintenance profile and minimal husbandry needs. That does not mean, however, that mice can simply be put in a habitat with food and water and never given any attention.

Almost all health problems in mice can be avoided by superior care, sound nutrition, and healthy interaction along with adequate intellectual stimulation.

Selecting a Habitat

You have a number of choices for a mouse habitat. Some people prefer wire cages for their superior ventilation and ease of cleaning, but you have to make sure there are no places in the cage where your pet can wriggle out or become caught.

Aquariums and Toppers

A standard 20-gal / 75.7 L aquarium works very well for 2 mice, but is somewhat harder to clean and must be monitored against overheating and a build-up of toxic ammonia.

A middle ground is to combine both concepts using a smaller aquarium, like a 10-gal / 37.8 L aquarium with an attached wire "topper," which snaps in place to create a more vertical habitat.

This will allow you to use internal structures like shelves and ramps to give your mice greater exercise opportunities in a more interesting setting.

Aquariums with toppers are often sold as "kits" and include some basic "furnishings." The Super Pet My First Home Tank Topper package includes a 10-gal / 37.8 L aquarium, topper, 3 shelves, 2 ramps, a food dish, water bottle, and nest box for $33 / £20.

(This unit can be purchased online from drsfostersmith.com as well as other retailers.)

A basic 20-gallon aquarium with a lid will cost approximately $40 / £24. A wire cage with a solid plastic bottom can cost anywhere from $35 to $100 (£21-£61) depending on its size and included features.

If you do opt for a wire cage, the solid base is essential or your mice can suffer from serious and painful sores on their feet. It's a mistake to have a cage with a wire bottom from a maintenance perspective anyway, since accumulated urine will cause the material to corrode.

The bars of any wire cage should be no farther apart than 0.25 in / 0.64 cm so that your mice cannot escape. This is roughly a size large enough for an adult to slip their index finger between the bars.

Habitrails

The "Habitrail" line of habitats made by the Hagen Corporation are almost iconic in the world of pet rodents. The modular design allows for the creation of elaborate tunnels and "warrens" intended to mimic a rodent's natural environment.

Over time, you can expand the structure, or simply rearrange the existing parts to make things more interesting both for you and your pets.

It's fun to watch pet mice (or hamsters, gerbils, or rats) running about in the "world" of a Habitrail. The components are well ventilated, but you will be faced with a slightly higher maintenance profile.

Everything snaps apart easily and can be washed, but it's important to make sure there is no build-up of either bacteria or ammonia. The entire habitat should be disassembled and thoroughly cleaned at least once a month.

To get started with a Habitrail set up, you will spend $35 to $50 (£21-£31) for the cage alone. The starter kit may or may not include food and water bowls.

Travel Carrier

In addition to a main habitat, I suggest you buy a small travel carrier for your mice. These units are just small plastic boxes with vented lids outfitted with handles.

Having a travel carrier will give you a safe place to secure your mice when you are cleaning their cage, or if a visit to the vet is required.

You must not leave your pets in a vehicle for any length of time unless either the air conditioning or heater is running. Mice will tolerate some degree of cold better than they will heat, which can prove to be deadly to the little animals very quickly.

Never allow the carrier to sit in direct sun even if the air conditioning is running. The sunlight will heat up the interior of the box very quickly, which can be deadly for your pets.

Clear plastic travel boxes with lids retail from $10-$20 / £6-£12 depending on size.

Choosing a Substrate

Two of the most popular options for substrate are ground corn cobs and shredded newspaper. Wood shavings or chips made of cedar or aspen are also an option, but their safety has been called into question in recent years.

Cedar, in particular, contains phenols that may cause health problems for small pets including liver disease and respiratory issues. Owners like cedar shavings because the aroma helps to mask the ammonia smell from urine.

If you plan to use cedar shavings, do so only in a well- ventilated cage, and consider putting a layer of corn cobs or newspapers on top of the cedar.

Newspaper is a safe option because most inks in use today are soya-based, but be sure to remove wet and soiled clumps of the paper daily.

Thankfully, mice tend to pick one corner to serve as their "bathroom," which makes spot cleaning easier for their keepers.

Kaytee makes a number of good substrates at reasonable prices:

Kaytee Kay-Kob Litter
8 lbs / 3.6 kg
$13 / £7.8

Kaytee Cedar Pet Bedding for Pet Cages
1000 cu in / 16 L
$11.79 / £7.15

Once a week, remove all the substrate and put down fresh material.

Temperature and Lighting

Regardless of the type of habitat you choose, do not put it in any area where it will be hit with direct sunlight during the day. Not only do mice prefer dim, indirect lighting, but they are easily subject to overheating.

In general, mice handle cold better than heat and should be kept at a temperature somewhere in the range of 65 - 80 °F / 18.3 - 26.7 °C.

I recommend installing a digital thermometer inside the cage. These units are readily available in pet stores and online since they are used by reptile keepers. A good example is Fluker's Digital Display Thermo-Hygrometer at $20 / £12.

While mice do not have specific humidity requirements, the measurement will help you to detect undue levels of moisture in the habitat that may signal the need to remove damp bedding and air out the enclosure.

You will be able to tell if your pets are overheated if they retreat to a corner and begin to hyperventilate. Mice do not try to drink more water when they are overheated, so they dehydrate rapidly and slip into a coma if the temperature is not quickly lowered.

Cage Maintenance

I've already mentioned the need for daily spot cleaning. The principle danger here is from ammonia fumes, which are unpleasant for you, but potentially deadly for your mice.

Ammonia irritates the lungs and will, over time, damage their lining. This makes your mouse more vulnerable to infections from viruses and bacteria. Cage size will actually help you to control ammonia levels.

The greater the surface area of the cage, the slower ammonia accumulates and the greater your chances of staying ahead of this development with spot cleaning.

Change out all the litter once a week and use this time as an opportunity to wipe out the base of the habitat with a 50/50 mixture of water and vinegar. Do the same with all habitat surfaces, making sure everything is dry before you return your pets to their home.

At least once a month, disassemble the habitat and take it outside or put it in the bathtub for a thorough washing. If possible, allow the habitat to dry in the sun.

If you are using a habitat with a plastic base, over time the material will absorb urine smells until you are no longer able to eradicate them with cleaning. At that point, you need to get your pets a new cage, or at least replace the bottom tray.

The Matter of Escapes

Mice are extremely territorial, so it won't take your pets long to regard their habitat as their domain. The greatest risk of escape is generally in the first few days your mice are in their new home.

Single mice are much more likely to try to get out because they're bored and lonely. Escapes are not much of an issue so long as mice are kept in pairs.

Diet and Nutrition

If you've ever had wild mice invade your storage pantry, you'll know that the little animals will eat virtually anything. This is not, however, the nutritional strategy you want to adopt with your pet mice.

Studies have determined that restricting a mouse's caloric intake will lengthen its life. Obese mice die much more quickly. Your best option is to purchase a pellet food especially formulated for rodents.

Examples of Pellet Foods

Pellet foods contain the right mixture of carbohydrates, protein, vitamins, and minerals for your pet. They are readily available and easy to use. Good examples include:

Mazuri Rodent Pellets

2 lbs / 0.9 kg

$6 / £3.6

Purina Garden Recipe Rat & Mouse Diet

4 lbs / 1.8 kg

$11 / £6.67

Kaytee Forti-Diet Pro Health - Mouse, Rat & Hamster

3 lbs / 1.4 kg

$5.79 / £3.5

It's best to avoid rodent foods that include seeds because of their lower quality and the high risk for fungal growth.

Since mice nibble constantly and need a constant food source, you can follow a "free feeding" program, keeping their bowl filled at all times.

Treats like garden vegetables, fruits, or nuts are fine so long as they do not comprise more than 20% of your pet's diet. Some acceptable vegetable options include:

- carrots
- corn on the cob
- bell peppers
- green beans
- squash
- sweet potato

Any kind of green leafy vegetables are also nutritious treats. Don't offer too many fruits, since they have such a high sugar content. Safe choices include:

- cherries
- melons
- bananas
- apples
- blueberries
- strawberries

Remember not to overfeed your pets. Obesity seriously shortens the lifespan of companion mice. If possible, only feed your pets organic produce, and wash all the items thoroughly so there is no risk of exposing your pet to herbicides of insecticides.

Any fruit or vegetable that you would peel before eating should also be peeled for your pet. Feed these items raw, so that the snacks also help to keep the mouse's teeth worn down. Do not give your pet any kind of canned vegetable or fruit since these items have high levels of sodium and may contain other, harmful chemicals, especially preservatives.

Foods to Avoid

Not all foods are good for your mice, and some should be avoided altogether. Do not give your pet oranges, lemons, or other citrus fruits since they may cause diarrhea. The same is true of garlic and onions, which carry the added risk of causing anemia.

Contrary to popular perception, it's best not to give your mice cheese since dairy products can cause gastrointestinal distress. Also, cheese is extremely high in fat and calories and can contribute to weight gain. Beyond all that? Most mice really don't even like cheese!

Under no circumstances should you ever allow your mice to have chocolate due to the presence of the alkaloid theobromine, which is toxic to many animal species including mice.

Avocado is also toxic to companion mice, and peanut butter presents a significant choking hazard. You can give your pets pecans and walnuts in the shell. The meats of these nuts are acceptable snacks, and the shells create a gnawing opportunity. (Offer nuts in moderation only.)

For all other food items, err on the side of caution and don't give them to your pet. It's never a good idea to get a companion animal started on "people" food. Mice that eat a diet of properly formulated pellets are much healthier and live longer.

Hydration

Provide your mice with clean, fresh, de-chlorinated water at all times. I recommend a water bottle that hangs on the side of the habitat with a ball bearing tip. Water dispensed in dishes is easily soiled with bedding and fecal matter, so bottles are a better option.

An 8-oz / 0.2-L hanging bottle with a ball bearing tip will cost approximately $5 / £3. Be sure to change the water daily and to test the ball bearing in the spout to make sure it is working well and dispensing adequate amounts of water.

Beyond your pet's physiological need for water, staying well hydrated will help your mouse to maintain its body temperature while flushing wastes and toxins from the animal's system.

Never allow your pet's water to become stale or stagnant since this creates a situation ripe for the development of bacterial growth.

Toys and Intellectual Stimulation

Mice are busy, busy, busy little creatures, constantly on the go and very curious. They love all kinds of toys and especially hollow objects like boxes and tunnels.

They will use wheels and ramps, and happily gnaw on old cardboard tubes and fiber egg cartons. Almost any chew toy rated for a bird or small rodent is fine for a mouse so long as it contains no potentially toxic substances or anything that might be a choking hazard.

Most toys that are suitable for mice are priced in the range of $3-$10 / £1.8-£6.

Be sure to include something that is a hard dense wood to give your mice something to really gnaw on. Remember that rodents' incisors grow throughout their lives and need to be kept worn down so your pet can eat properly.

Definitely include a small running wheel with a solid plastic track so your pet's feet won't get caught in the rungs. The smallest size, which will be around 4.5 in / 11.4 cm costs approximately $8 / £5.

(A good "do-it-yourself" option to create tunnels and mazes for your pets is simply to get lengths and joints of PVC pipe. You won't be able to see your mice inside the pipes, but this is a less costly alternative to having a "Habitrail" enclosure.)

Bathing

Mice don't need to be bathed the way other companion animals do, and certainly not in water. If you want to provide your mouse with a real grooming treat, offer him a dust bath.

Not only do mice love rolling around in a dish of dust, but the process removes excess oils and dirt from their coats – and it's great fun to watch!

You can use a product like Chinchilla Dust Bath (2.5 lbs / 1.1 kg), which costs approximately $10 / £6. Just put a small amount in a shallow bowl and offer it to your mice. They'll know what to do from there.

Always remove the bowl when your pets are done. Do this once a week, and let your mice have 20-30 minutes to use the dust. Don't leave the material in the cage all the time, or your pets may develop eye irritation.

Since this can create a lot of mess, I think it's a good idea to offer your pets their dust bath on a day when you're planning to clean the entire habitat anyway. Let them have a good roll in their dust, and then remove them to their travel box while you put their house back in order.

Handling

For the first 2-3 days after you bring your mice home, resist the urge to handle them too much. Let them get used to their new surroundings before getting really acquainted with your pets.

Don't stand over the cage looking down at them. This kind of "looming" or "towering" behavior is threatening, especially when mice are transitioning to a new home.

When you think the time is right, put your hand in the habitat and let the mice come to you for a good sniff. Mice don't see very well, and primarily interpret and interact with their environment through the sense of smell.

In many cases, well socialized mice will just climb right up into your outstretched palm. The correct way to pick up a mouse is to gently grasp your pet at the base of the tail and then slide the animal into your cupped hand. Always be very gentle, as mice are fragile.

Speak gently to your mice. They have acute hearing and a keen sense of vibration. Loud noises frighten them, and their whiskers are sensitive to sudden movements of the air. Move slowly.

If you really need to pick up your pet or move it to its travel crate and the mouse is not cooperating, gently nudge it into a toilet paper roll and then cover both ends with your hands.

Little bits of treats are always helpful in encouraging your mouse to come to you and to be handled. Domesticated mice are rarely aggressive or even temperamental. They learn quickly and will soon see you as their friend and keeper.

Training Your Mice

There is a good reason why mice are so frequently used by scientists performing studies on behavior and psychological responses. Mice are very smart, and are adept at problem solving.

Obviously your mouse isn't going to be playing fetch with you, but there are many creative ways you can elicit desired responses from your pet. Think about "training" more in terms of improved communication.

Much of the process is predicated on observing the things your mouse is already doing and reinforcing behaviors that you find desirable with an eye toward extrapolating those actions into more complex sequences.

Any time your mouse does something you want him to do or that you like, always give him a treat. Although some people do use clickers to train their mice (small handheld devices that make a clicking noise when triggered), I think that edible rewards are the best option for pet owners.

Pair the treat with praise. In time, your mouse may come to respond simply to the pleased tone in your voice. Mice are affectionate, and they do have a desire to please their humans.

Never speak loudly to your pet. The only "punishment" for not performing a task is not getting a treat or not being praised. Ignore incorrect behavior and reinforce correct behavior. That's the formula.

There is no set regimen of tricks. If your mouse naturally scampers up a rope in the cage, and you would like him to do that when you tap on the cage top, or say a particular word, reinforce for that sequence of events.

Use your imagination, think outside of the box, and cater to your mouse's natural interests and talents. Remember, this is a game. Treat it that way, and make it fun -- for you both.

Chapter 5 - Estimated Costs

This list of costs considers the purchase of a mouse and the basic set up and supplies. Obviously some of these expenses are ongoing per month.

Cost of Your Mouse

Feeder Mice

$0.30-$0.50 each / £0.18-£0.30 each

Fancy Mice

$10-$30+ / £6-£18+

Habitat Choices

aquarium with topper and accessories

10 gal / 37.8 L

$33 / £20

20 gallon aquarium

$40 / £24

wire cage with a solid plastic bottom

$35 to $100 (£21-£61)

Habitrail line of habitats

"starter" cage

$35 to $50 (£21-£31)

clear plastic travel boxes

$10-$20 / £6-£12

Substrates

Kaytee Kay-Kob Litter

8 lbs / 3.6 kg

$13 / £7.8

Kaytee Cedar Pet Bedding for Pet Cages

1000 cu in / 16 L

$11.79 / £7.15

Miscellaneous

digital thermometer with hygrometer

$20 / £12

dust bath

2.5 lbs / 1.1 kg

$10 / £6

Foods

Mazuri Rodent Pellets

2 lbs / 0.9 kg

$6 / £3.6

Purina Garden Recipe Rat & Mouse Diet

4 lbs / 1.8 kg

$11 / £6.67

Kaytee Forti-Diet Pro Health - Mouse, Rat & Hamster

3 lbs / 1.4 kg

$5.79 / £3.5

Water Bottle

8 oz / 0.24 L hanging bottle
with a ball bearing tip
$5 / £3.

Toys

chew toys
$3-$10 / £1.8-£6.

exercise wheel
4.5 in / 11.43 cm
$8 / £5.

Estimated Set-Up:

The following figures are highly subjective and suggest only a high-low range based on the minimum required supplies listed above

$95 - $200
£58 - £122

Chapter 6 - Health and Breeding

If you are considering keeping a pet mouse, you must understand that your pet won't be with you for the same length of time you'd enjoy with a more conventional animal companion.

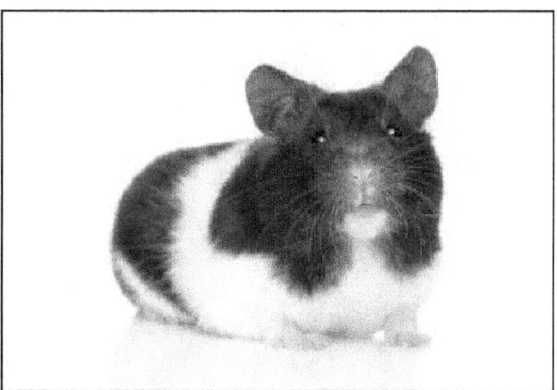

With very few exceptions, like a respiratory infection or external parasites, there's little that can be done to treat mice with more serious conditions. The one notable exception is broken bones, which can be splinted by a veterinarian and in most cases heal quite well.

Overall, however, mice are small, fragile creatures who will benefit most from preventive health care in the form of excellent husbandry, a responsibility that falls squarely with you.

Observe Your Mice Carefully

The cornerstone of good health care is observation. Only you will know what is normal for your mice, so the abnormal must be evaluated as a potential problem. All of the following can be signs of physical illness:

- Any change in weight, especially weight loss.
- Reluctance to move or difficulty moving.

- Limping or an otherwise uneven gait.
- A scruffy or unkept look to the mouse's coat.
- Frequent scratching to the point of physical injury.
- Sitting in a hunched position with the tummy tucked up.
- Chattering, sneezing, or raspy breathing.
- Any discharge or crusty accumulation at the nostrils.
- Soiling around the anus or poorly formed stools.
- Any discoloration of the skin.
- Open abrasions and wounds.
- Tooth grinding.
- Any growths, bumps, or lumps.
- Distention of the abdomen.
- Clouding of the eye, squinting, or discharge.

Some of the problems with husbandry that take a tremendous health toll include over-crowding, poor diet, and a build-up of ammonia fumes from soiled litter.

Common Illnesses and Problems

The following issues are the ones most commonly seen in pet mice. Some can be improved by changes in husbandry practices, and others by veterinary care if you have a vet willing to treat "pocket pets." Others, however, can only be addressed by kind, palliative care.

Respiratory Infections

Respiratory infections are a leading cause of death in pet mice. If your mouse is chattering constantly and potentially wheezing, an infection is likely the cause.

If you have more than one mouse, remove the affected individual immediately and place your pet in quarantine in an effort to protect the other mice in the habitat.

Mice do respond well to antibiotics obtained from a veterinarian and administered orally through an eyedropper or with a syringe with the needle removed.

If you suspect your mouse has a respiratory infection, consult with your vet about the possibility of acquiring medication.

One of the best ways to guard against your mice developing respiratory problems is to spot clean the cage daily with full litter changes each week.

Mice typically pick one corner of the cage where they urinate. It's important not to leave the soiled bedding in place long enough for levels of ammonia to build-up in the habitat.

The ammonia damages the mouse's lungs and makes the animal much more susceptible to infection.

Excessive Scratching

Usually when a mouse scratches all the time, even to the point of injuring itself, there is some type of external parasite present, often mites or lice.

You can treat your pets with an anti-parasitic like ivermectin or pyrethrin, but you must be sure of the strength and dosage. Always consult with your veterinarian or with a knowledgeable mouse breeder about how to administer this kind of treatment.

If you suspect that parasites are an issue, you will also need to thoroughly clean and disinfect the mouse's habitat and provide your pet with completely fresh bedding.

If your mouse is scratching so vigorously that its skin or ears are developing open wounds, try "gloving" your pet's back feet with tiny pieces of masking tape folded over the paws. This will allow your mouse to get around and play normally, but prevent your pet from injuring itself.

External parasites may take 2-3 weeks to resolve. During that time, you will likely need to retreat your pet with the recommended preparation and clean the cage multiple times to ensure the parasites do not reproduce.

Diarrhea

Diarrhea in pet mice can be caused by a number of problems including the use of antibiotics for another illness. The antibiotics destroy the flora in the mouse's gut leading to an overgrowth of bacteria.

If the diarrhea has been caused by an antibiotic, the condition can be cured with the use of a probiotic like acidophilus added to your pet's food.

Another explanation might be an electrolyte imbalance from improper diet. In that case, a veterinarian could prescribe a fluid supplement. If anything new or unusual has been added to your pet's diet, that item should be discontinued immediately.

Growths and Lumps

In almost all cases, a growth or lump on a pet mouse will be either a tumor or an abscess. Most vets will first attempt to determine if the growth is an abscess, which can be drained and then treated with an antibiotic.

If the lump does not begin to shrink in a few days or refills with fluid, the vet will aspirate the mass to test for cancer. The most common tumor in pet mice and rats is on the mammary gland with metastasis to the lungs.

In some cases owners will opt to treat a cancer with a drug like tamoxifen, which may add 2-3 months to the animal's lifespan. If untreated, tumors are generally fatal in 2-3 weeks.

For most pet owners, treating cancer in mice is not an option due to the expense. The measures are sometimes taken with show mice, but it's best to follow your vet's advice regarding the most humane course of action.

Ringworm

Ringworm or dermatophytosis is a fungal skin infection. If your mouse develops ringworm, the animal will scratch excessively and may develop open, crusty wounds.

The fungus will need to be cultured by your veterinarian to get an accurate diagnosis, but it can then be treated fairly easily with a topical medication. Until the ringworm is resolved, wear gloves while handling your pets.

Swelling of the Abdomen

If abdominal swelling is present in a female, pregnancy must first be excluded. In all other cases, the swelling is indicative of an internal cancer, most likely lymphoma or leukemia.

The mouse may also exhibit shallow breathing and there may be swelling of the lymph nodes. Very little can be done to help the mouse, and since the little animal will be in a great deal of pain, euthanasia is recommended.

Old Age

As mice age, they develop problems for which there really is no practical solution. In addition to tumors, elderly mice are highly susceptible to kidney failure.

It is also possible for dental problems to create a misalignment of the animal's upper and lower incisors, which prevents the mouse from eating appropriately.

As a mouse ages, the most important things you can do are to ensure that your pet is still getting adequate food and water and has a place to curl up and stay warm.

Given the short lifespan of pet mice, watching this process is an inevitable consequence of being a mouse owner. Keeping your pet comfortable and loving the little animal to the end of its life is the best you can do.

The Issue of Zoonotic Diseases

Zoonotic disease are illnesses that can be passed from an animal back to a human. This is not a worry with domesticated fancy mice, but wild mice can be the source of illnesses in humans.

Colonies of wild mice can harbor a number of viruses and bacteria that are dangerous to humans including, but not limited to:

- salmonella

- lymphocytic choriomeningitis virus (LCM)

- leptospira

- giardia

- cryptosporidia

- various skin fungi

Certainly, pet mice are not immune to also harboring these microorganisms, but the prevalence of these pathogens in wild mice is due to the highly unsanitary conditions of large mice colonies.

As a survival tactic, these colonies are often located in enclosed areas where feces and urine build up in great quantities. It is for this reason, as well as for the nuisance factor, that infestations of wild mice in homes and other buildings must be eradicated.

The potential for disease transmission to children, the elderly, and those with compromised immune systems is too great to allow a colony of mice to stay in place.

Breeding Mice

Female mice reach sexual maturity at 6-7 weeks of age, with males following suit about one week later. Mice should not be bred, however, until they are at least 3 months of age. Females should not be allowed to become pregnant after they are one year old.

Fertility in Females

Female mice cycle every 4-5 days if they have not been impregnated. They are then fertile for approximately 12 hours. Breeding typically occurs at night and can happen at any time of the year.

In planning the breeding of mice, pick your breeding pair carefully. You not only want healthy adults, but also those with a good temperament.

Determining that Mating Occurred

You may or may not know if your mice have mated. The best way to check after a pair have been placed together is to check for the presence of a vaginal plug in the female.

The vaginal plug is cream colored and waxy in appearance and will drop out within 24 hours. You can gently examine the female or look for the discarded plug in the habitat, although it may be difficult to find depending on the type of bedding used.

Pregnancy and Birth

Pregnancies last 18-21 days with litters typically numbering 4-15 pups. A female can have 5-10 litters if allowed to do so in a single 12-month period.

Baby mice weigh, on average, 0.5-1.5 grams. As a point of comparison, it would take 4 mouse pups to equal the weight of a single cube of sugar.

The pups are born without hair. They are bright pink and their eyes are closed for the first 10-14 days of life.

During the first 2-3 days after birth, a mother mouse should not be disturbed. If females become upset, they may eat their young.

Nursing continues for three weeks, but the female is fertile again for 14-24 hours after birth. After that timeframe passes, the female will not be fertile again until after she has weaned her young.

If two female mice are kept together, they will often not have their usual cycles of fertility. If a mature male is introduced to the habitat, however, both females will resume cycling in a few days.

Understanding Genetics

If you are breeding your mice for specific traits like color and coat texture, you will have to acquire an understanding of the genetics involved.

This is an extensive topic and one on which whole books have been written. The basic issue is understanding dominant and recessive genes and then calculating the possible outcomes of their pairing.

Genetics for color or pattern are not the only consideration, however. If you have a mouse that is unusually aggressive or has an otherwise undesirable behavioral trait, you will want to isolate that individual to prevent the trait from being passed on.

A good place to begin your education in mouse breeding and genetics is the Fancy Mice Breeders Forum at FancyMiceBreeders.com.

Any time I suggest that my readers become involved in online discussions, I like to offer a word or two about etiquette in these types of venues. It's important to understand a little bit about the culture and etiquette of forums before you just jump in.

Typically, forums have a special area set aside for newcomers to introduce themselves. After you do that, spend some time "lurking." This just means reading the discussion threads and getting a feel for how people in the community interact with one another.

Since the regulars on forums come to know one another quite well, it's easy for newcomers to give offense without meaning to simply because they don't understand the dominant forum culture.

Should You Breed Mice?

Obviously, only you can answer that question, but given the frequency with which females are fertile and the size of their litters, think carefully before allowing your mice to mate.

Overcrowding of mice is cruel and a sure recipe for the spread of illness. The more mice you keep, the harder it will be to maintain a good standard of husbandry for their habitats and the more work you are creating for yourself.

If you buy your pet mice from a breeder, there should be no risk of unplanned litters since accurate sexing of the pair allows you to choose female/female cage mates.

Only breed mice if you have the room to care for them properly or if you have placed the pups in advance. It's far too easy to suddenly find yourself with 50 or 100 mice that take up all your time and attention.

Chapter 7 - Rodent Shows

Many people are surprised to find out that companion rodents are exhibited in shows, but the first of these events actually date to the early 1900s in England.

The shows not only promote the keeping of mice and rats as pets, they also give breeders a place to show off their animals and to compete for awards.

How Rodent Shows Operate

Rodent shows are very similar in operation to those held for the exhibition of any other type of animal. The event is sponsored by a governing organization and entrants are judged according to recognized standards.

Breeders work with their bloodlines to produce animals that conform as closely as possible to those standards, with the finest examples competing for the "best in show" titles.

Shows will include a wide variety of "classes" to maximize participation, including those specifically organized for young people just getting involved with the hobby.

American Fancy Rat and Mouse Association

The majority of exhibits held in the United States are under the oversight of the American Fancy Rat and Mouse Association (AFRMA) with the following requirements:

- All animals must be shown by their owner or by a responsible party.

- The organization will take all reasonable care to ensure the safety of the animals, but cannot be held responsible for loss, death, or injury.

- All animals must be shown in their natural condition without any alteration to their coat or color in any form.

- Entry fees are non-refundable.

- All animals must be judged as entered even if they have been placed in the wrong class and are subject to elimination.

- There is to be no discussion between exhibitors and judges before or during judging unless the judge initiates the contact.

- All placements made by the judges are final.

- Exhibitors are expected to set good examples of co-operation and sportsmanship.

All animals exhibited are judged according to the Official Standards of the American Fancy Rat and Mouse Association.

AFRMA Mouse Varieties

The following are the seven varieties of fancy mice recognized by the AFRMA by coat type:

- Standard: A coat that is short and sleek.

- Satin: A short, sleek coat with a lustrous sheen.

- Long Hair: Hair that is thick, long, fine, and silky.

- Long Hair Satin: Hair that is thick, long, fine, and silky with a satin sheen.

- Frizzie: Dense and tightly frizzed or waved hair covering the whole body with curled whiskers.

- Frizzie Satin: A coat that is frizzed or tightly waved that also has a satin sheen. The mouse should also have curly whiskers.

- Hairless: Mice that are thin, with bright, translucent pink skin and no pimples or scars. These specimens should have no hair whatsoever.

Each variety is then grouped into five sections by color and markings:

- Self mice are of a uniform color in black, beige, blue, chocolate, coffee, cream, champagne, dove, gold, fawn, lilac, ivory, red, orange, silver, or white.

- Tan and fox mice have a top color in any recognized shade with an underside that contrasts. If the underside is a rich golden tan with red tones, the mouse is a "tan."

If the underside is almost white, the mouse is a "fox." The line between the top and bottom colors should be clear, distinct, and straight along the jaws and chest.

There are seven groups in the "marked" section that may be present in any recognized color. (Please see Chapter 2 for a more complete explanation of markings as this can be a complex and confusing topic to summarize.)

There are also categories for unstandardized animals that have not yet been accepted or recognized by the governing body as well as those for non-recognized colors.

General Characteristics of Show Mice

In general, show mice should have long, slender bodies, with large, clear eyes. Their heads should have clean lines in profile as well as in width and their ears should be long and expressive. Long tapering tails are valued, with the average overall length coming in at 8-9 in /20.3-22.9 cm.

The mice should be easy to handle and have a good temperament. Mice are judged on type, color, markings, condition, coat, head, ears, eyes, tail, and size.

The National Mouse Club

The National Mouse Club in the United Kingdom has similar standards for its shows which are published on the group's website at TheNationalMouseClub.co.uk.

The club's general standard of excellence calls for a mouse to:

- be long on body with a long clean head
- have a nose that is not too pointed or fine
- have eyes that are bold, large, and prominent
- have tulip shaped ears that are large and free of creases
- carry its ears erect with plenty of width between

The body should be slim and long with a hint of arching over the loin for an appearance that is "racy." The tail must be straight and free of kinks, thick at the root, and tapered like a whiplash to a fine point. The tail should be roughly the same length as the body. Unless the variety calls for a different texture, the coat should be sleek, glossy, and smooth.

Chapter 8 - Frequently Asked Questions

While the main text of this book is intended to fully discuss everything you need to know about keeping mice as companion pets, the following are some common questions asked about mouse husbandry.

How do mice behave as pets?

Your mice will be most active early in the morning, just as the sun is going down, and throughout the night. Mice very rarely just sit around doing nothing. They are constantly on the go, checking things out, eating, grooming, and just in general attending to the business of being a mouse.

Initially, your pet will be cautious in his interactions with you, which is understandable given how enormous you must look to the little animal. Don't worry, however, mice become socialized very quickly.

They are highly intelligent, recognizing and responding to kindness and gentleness. It won't be long (generally just 2-3 days) before your pets are greeting you enthusiastically and happily climbing into your outstretched hand.

Can I tame a wild mouse?

I'm certainly not going to say it's impossible to make a pet of a wild mouse, but it's extremely difficult to do so. Wild mice, sometimes referred to as "house mice," never really settle down in captivity.

The best pets are domesticated mice. You can either buy the mice sold in pets stores as "feeders" for larger animals or get a "fancy" mouse from an established breeder.

Do feeder mice make good pets?

Feeder mice can make good pets, but they are typically bred in deplorable conditions with little to no care taken about their health since their fate is assured from the moment of birth.

Many prospective mouse owners take pity on feeder mice and regard the adoption as a rescue. With love and attention, feeder mice can be lovely little pets, but they are never as healthy or long lived as fancy mice.

Why are fancy mice better pets?

Fancy mice acquired from breeders are better pets because they are the product of planned breeding programs. They have received excellent care, and they come to their new owners well socialized.

There is also a much greater variety of coat types, colorations, and markings present in fancy mice, making them more visually attractive as pets.

How many mice should I get?

If possible, you always want to get two mice. They are social animals and can get bored, depressed, and even aggressive when housed alone. Also, single mice tend to work harder at escaping their habitat because they're looking for something to do.

Should I get males or females?

Your best option is to adopt two females of the same age and size. Males can be housed together if they are from the same litter and have never been separated, but you may still see some territorial aggression.

You might not believe it given their small size, but when two mice fight, it can be a vicious and bloody battle.

How are genders distinguished?

If the mice are young, you have to look at the two anal openings. In male mice, both openings are round. In females, the opening nearest to the body is a small slit about the size of a hyphen.

As the mice get older, however, a male's scrotum and testicles are clearly visible, making identification quite simple.

How do I introduce a new mouse?

It may take two or three days for two adult mice to work out the pecking order of their new living arrangement. It's important to observe the animals during this period and to make sure there's no serious fighting.

You won't really know that the two have worked things out until you find them sleeping together in the nest box. Once that happens, the arrangement has been solidified.

The important thing, however, is to quarantine the first mouse in another enclosure for at least 3 weeks to make sure the new animal is healthy before you put it in the main habitat with your existing pet.

Are pet mice hard work?

On average, most mice don't live past 2.5 years, and many die in the first year of life. During that time you obviously must commit to feeding and caring for your mice by maintaining their habitat and interacting with them.

On a daily basis their habitat should be spot cleaned, with all the litter changed out once a week.

Mice are extremely clean in their habits, and do best on pellet foods with the occasional treat, so on a whole, no, they are not labor intensive.

Do not, however, make the mistake of thinking your mice don't need interaction with you. Like any pet, mice also require an emotional commitment from their humans.

So what do mice eat?

In truth, a mouse will eat just about anything, but that's the worst possible approach to nutrition you can take on behalf of your pet.

Studies have found that mice that are fed a well-regulated diet and that are not allowed to become obese, live longer.

The staple of your pet's diet will be a standardized pellet food formulated for rodents to contain the correct mixture of protein, carbohydrates, vitamins, minerals, and fiber.

This can be augmented with the occasional treat, a topic fully discussed in the chapter on daily care. (Please note that the same material also references items you should not feed to your pet.)

How much should I feed my mice?

Mice are nibblers by nature, so you will want to "free feed," meaning they should have a constant supply of food pellets in their bowl.

How much water do mice need?

A constant supply of clean, fresh water is a must for pet mice. I recommend a bottle that hangs on the side of the habitat and has a ball bearing tip. This prevents debris from getting in the water and is easily changed each day without any spillage.

What kind of cage should I get?

The primary considerations with a habitat are roominess, warmth, and security. Most people use aquariums. A 20 gal / 75.7 L aquarium with a screened lid will work very well for 2-3 mice.

Some people think wire cages offer better ventilation and are easier to clean, but you have to be very careful to get a cage that does not have places for your mice to catch their feet or legs and potentially break a bone. Also, escapes are more likely with wire cages.

The major drawback with aquariums is that the litter must be spot cleaned daily and completely changed weekly to guard against a buildup of ammonia from accumulations of urine.

It's also important to monitor the interior temperature of an aquarium to make sure it stays in a range of 65-80 °F / 18.3-26.7 °C. For this reason, I always recommend the use of a thermometer attached to the side of the habitat.

No mouse habitat should ever be placed in direct sunlight, not only due to the danger of overheating, but also because mice prefer indirect to dim lighting.

What are some good toys for mice?

A running wheel is an absolute must for these incredibly active little animals. Always get a wheel with a solid plastic track, not rungs.

Beyond that, your pets will like anything on which they can climb or into which they can burrow. They also enjoy chewing and shredding.

You can recycle some common household items for your little pets to destroy, like cardboard tubes from toilet tissue or paper towel and fiber egg cartons (not Styrofoam.)

Most toys designed to be gnawed on by either birds or other rodents are fine for mice, just make sure there is nothing toxic on the toy and that there are no choking hazards.

Sisal and hemp rope toys in the habitat are great for climbing, and most mice really enjoy wicker and straw baskets as well as any kind of tunnel.

Some habitats are designed to incorporate elaborate tunnel systems that can be augmented or rearranged to keep things interesting for your pets.

It's actually a great deal of fun to design and outfit a mouse habitat because it's so entertaining to watch the little animals busily going about their daily "business."

Is my mouse sick?

Basically when a mice stops moving all the time, something is wrong. Other indicators of ill health include a hunched over posture, weight loss, and deterioration of the coat. Healthy mice are glossy and sleek, with smooth coats (unless you have a fuzzy or rex breed with wavy hair.)

If you think something is wrong, examine your mouse gently looking for wounds, possibly broken appendages, or any abnormal growths.

Many veterinarians will examine and attempt to treat pet mice, but often there is little that can be done for your pet but to make it comfortable and to keep it warm.

Sadly, these delightful little animals do not live more than a year or two. I think it's especially important to consider this fact when thinking about a mouse as a potential pet for a child.

Is chattering normal?

Chattering is usually a symptom of a respiratory illness, which is a leading cause of death in pet mice. Oftentimes, mice can be treated with antibiotics administered orally with an eyedropper or a syringe with the needle removed.

Consult with your veterinarian. If you have more than one mouse, isolate the individual that seems ill in an effort to stop the spread of the infection.

Deciding to euthanize?

If you have an older mouse that is refusing to eat and seems to be suffering, call your veterinarian. The kindest way to help a terminally ill mouse is euthanasia. There is no humane way to do this at home, but your vet will be able to help.

Chapter 9 - Care Summary

Although the entirety of this text is dedicated to explaining mouse husbandry, the following is intended as a quick reference guide.

General

Mice are small rodents with pointed snouts, rounded ears, and hairless tails. Their incisors, which are used for gnawing and self-defense grow throughout their lives. But companion mice rarely, if ever, bite.

Cost for fancy mice: $10-$30 / £6-£18

Picking Your Pet

There is no preference for selecting pet mice by gender, although females do better when housed in pairs. Males have a tendency to become territorial and may show aggression toward one another. If males are to be kept together, they should be littermates that have never been separated.

Mouse Hygiene

Mice are very clean and groom themselves just as cats do. They pick one corner of their habitat to urinate. Mice are not hypoallergenic, and can cause a strong allergic reaction in some people.

Signs of Ill Health

In selecting a pet mouse, all of the following are indications of ill health:

- sneezing, wheezing, or rattling respiration
- thin physical condition
- discharge of the eyes or nose
- soiled rump

- hunched posture
- bloated belly
- ragged or tacky coat
- depressed attitude and lack of interest

Life Span and Size

1-2.5 years
1-2 oz (28.3-56.7 g)

Habitat

Mice will thrive in simple, gnaw-proof habitats with screened lids or other means of security to prevent escape and to protect the mice from other household pets like cats.

The recommended habitat size for two mice is the spatial equivalent of a 20 gal / 75.7 L aquarium.

Bedding

The recommended bedding types are ground corn cobs or shredded newspaper. Although some people use cedar shavings, this is strongly discouraged as the phenols in the wood are harmful to many small animals including mice. Both respiratory distress and liver disease can result.

Recommended Accessories

- water bottle with a ball bearing tip
- food bowls
- nest box with soft, shreddable bedding
- exercise wheel
- toys including tunnels, ladders, and chew toys

Nutrition and Hydration

Nutritionally balanced mouse pellet foods and occasional treats of fruits and vegetables. Mice should have a constant supply of clean, fresh water changed daily.

Maintenance

Spot cleaning daily, complete litter changes weekly, thorough habitat washing and airing monthly.

Chapter 10 - Mice in Popular Culture

Mice are so delightful and whimsical by nature, they have found their way into many popular culture references. Of course, the most famous mouse in the world, Mickey Mouse, the beloved 1928 Walt Disney creation bears no true resemblance to the real thing.

In the world of literature, however, our clever little pets have been unleashed to have the kind of adventures of which we think they might be capable as we watch them busily going about their affairs.

If you've never really watched a mouse, you may not realize just how important he can seem!

Literary Mice

I always like to talk about literary mice in the hopes that young enthusiasts will not only love these exceptional little animals, but also fall in love with reading.

My first introduction to a mouse on the printed page was in a Newberry Award-winning book by Robert C. O'Brien, "Mrs. Frisby and the Rats of NIMH."

Mrs. Frisby is a valiant mother mouse protecting her children from a farmer's plow aided by the super-intelligent rats of NIMH. The rats, laboratory escapees, help her save her children while she helps them avoid recapture.

This is a book about loyalty, bravery, friendship, and family -- and it really made me want a pet mouse, much to my mother's horror!

A whole new generation of children became acquainted with Stuart Little, the creation of writer E.B. White in the 1999 film starring Michael J. Fox and Geena Davis.

The story is about a family accepting the unexpected birth of a son who just also happens to be a mouse. Stuart lives one adventure after another in spite of his tiny size.

Stuart teaches us all lessons about love and tolerance in this unlikely interspecies tale that happily invites readers to suspend disbelief and fall thoroughly in love with a winsome creature with a huge heart.

The Redwall Series by Brian Jacques takes us into a sprawling world that develops through 22 novels all set at Redwall Abbey where the mice have an intricate social and political system full of intrigue and true heroes.

The first book in this epic children's fantasy series was published in 1986 and is called simply "Redwall" -- and be warned, you'll be immediately hooked.

Mice on the Silver Screen

Mickey Mouse made his entrance into our hearts in 1928 in the animated short "Steamboat Willie." Then he was seen in black and white, but today we know him for his white gloves, red shorts, and bright yellow shoes.

As the official mascot of The Walt Disney Company, Mickey is an ebullient ambassador for the "happiest place on earth."

He made possible characters like Templeton, the mouse who steals the show on "Charlotte's Web" (1973), the courageous mice duo in "The Rescuers" (1988), and Remy, the mouse with dreams of being a world class chef in the 2007 animated film "Ratatouille."

And no one who loves the classic Disney cartoons can forget the mice who sing to Cinderella as she goes about her chores and endures the taunts of her stepmother and stepsisters.

Real Mice and their Music

As I was researching this book, I was delighted to find an article from *Time* magazine by Sonia van Gilder Cooke from October 2012. It seems that mice do sing.

Male mice serenade their ladies as part of the courtship ritual, but studies conducted at Duke University and published in the journal *Plos One* revealed some remarkable details about this behavior.

Male mice have the ability to sing in unison, in the same key, on pitch, and they can modify and adapt their melodies.

When they are singing, the mice are using the same part of their brains that humans use when we create music. Unfortunately, mice songs fall in an ultrasonic range the human ear can't detect, but that doesn't make this knowledge any less charming.

I found a modified audio file online of mice singing. The tones were changed to bring them into the range of human hearing.

The mice may not be little Pavarotti imposters, but the modulation was clear and they were definitely singing in unison -- a fact not lost on my cats who came trotting in from the back of the house with a look of intense interest on their faces.

Perhaps one of the reasons we are so intrigued by mice in life and in popular culture is that though they are very unlike us, we still have common ground with these tiny creatures that are so willing to share their lives with us.

Breeders – United States

Connecticut

The House of Mouse
http://www.freewebs.com/thehouseofmouse

Indiana

What the Cat Dragged In
http://www.whatcat.wordpress.com

Kansas

Little Loves Mousery
http://www.facebook.com/pages/Little-Loves-Mousery/444716415539380

Kentucky

Jack's Mousery
http://www.jacksmousery.com

Maryland

Mason Dixon Rodentry
http://www.masondixonrodents.com

Michigan

Silver Fuzz Rattery
http://www.silverfuzzrattery.com

New Hampshire

Storybook Pocket Pets
http://www.freewebs.com/storybookspocketpets/index.htm

New Jersey

Darby Mousery
http://www.darbymousery.wordpress.com

Ohio

Mousykins
http://www.mousykins.info

Oregon

Runaway Mousery
http://www.runawaymousery.weebly.com

Pennsylvania

CS Beck Rodentry
http://www.csbeck.com

Texas

Martingale Mousery
http://www.freewebs.com/martingalemousery

Washington

Twitching Whiskers Mousery
http://www.twitchingwhiskersmousery.com

Breeders – United Kingdom

Faluma Mousery
http://mousery.faluma.us

Jingles Mousery
http://www.sites.google.com/site/jinglesmousery

Wolf Magic Rattery
http://www.freewebs.com/wolfmagicrattery

Bumblebee Mice Mousery
http://www.bumblebeemice.synthasite.com

Painted Skies Mousery
http://www.paintedskiesmousery.webs.com

Breeders – Australia

Jaroslava Mousery
jaroslavamousery.webs.com

Mousekateers Mousery
http://www.freewebs.com/mousekateers

The Sweet Mousery
http://www.thesweetmousery.tripod.com

Captain Ratz Mischief
http://www.captainratz.com

Breeders – Canada

Rocky Moutain Mousery
http://www.rockymountainmousery.webs.com

Fancy Fuzz Mousery
http://www.fancyfuzzmousery.webs.com

Have Mice Will Travel
http://www.havemicewilltravel.homestead.com/index.html

Sputnik Mousery
http://www.sites.google.com/site/sputnikmousery

Afterword

If you recall the story I told in the foreword about my friend Larry, you'll understand what I mean when I say I don't think he ran right out and got himself a fancy mouse after his experience with a wild one.

However, being "invaded" by a tiny brown field mouse did change his appreciation for mice in general — for their cleverness and their adaptability to living in proximity to humans, whether they've been invited or not.

This book is about mice that are invited into their human's lives. Fancy mice come in all colors, coat types, and patterns. They are the product of planned breeding programs and are even exhibited competitively in rodent shows.

Mice are attractive as pets because they are inexpensive and easy to look after, but also because they are intelligent and inquisitive. With very little "training" mice become accustomed to being handled and appear to enjoy the interaction.

When given habitats with lots of furnishings and tunnels, their busy comings and goings are fascinating and fun to watch, not just for children, but also for adults.

Do not, however, think that you can just put mice in their habitat and ignore them. Like all companion animals, mice need and want time with their humans. Mice can become bored, lonely, and even depressed.

They are fastidious little creatures that need good husbandry to stay healthy and to live out their life span of 1.5 to 2 years. Something as simple as a buildup of ammonia fumes from wet litter can seriously harm your pet's health.

If there is any major "con" to keeping pet mice, it's that they don't live long. If you are getting mice for your son or daughter, be prepared for this fact, and gently prepare your child for this reality as well.

You may be surprised how attached you can become to a mouse, and how hard it is to say good bye to your pet when that eventuality finally occurs.

If you can handle their short lifespan, and the need to spot clean their habitat daily, there is little else complicated or difficult about keeping mice as pets.

They thrive on pellet food and the occasional snack of fruit or vegetable and will get on perfectly well with shredded newspaper for litter, playing their tiny role in household recycling.

In my experience, there is little to actually "go awry" with pet mice. For people with limited space who cannot keep a dog or a cat, pet mice are actually a well laid scheme for delightful companionship in a compact but thoroughly adorable package.

Relevant Websites

American Fancy Rat and Mouse Association
http://www.afrma.org

The History of Fancy Mice
http://www.afrma.org/historymse.htm

Rat and Mouse Club of America
http://www.rmca.org

Fancy Mouse Breeders' Association
mousebreeders.org

Fancy Mice Breeders' Forum
http://www.fancymicebreeders.com

East Cost Mouse Association
http://www.eastcoastmice.org

The National Mouse Club
http://www.thenationalmouseclub.co.uk

The London & Southern Counties
Mouse and Rat Club
http://www.miceandrats.com

South Eastern Mouse Club
semouse.webstarts.com

The Scottish Mouse Club
thescottishmouseclub.webs.com

Australian National Rodent Association
anraq.tripod.com

Australian Rodent Fanciers' Society of NSW, Inc.
ausrfsnsw.com

Australian Rodent Fanciers' Society Qld. Inc.
http://www.ausrfsqld.com

Russian Mouse Fanciers
miceland.ru

The Finnish Show and Pet Mice Club
http://www.hiiret.fi

Netherlands Mouse Club
http://www.kleineknaagdieren.nl

Polish Mouse Club
http://www.pmcorg.webd.pl

Deutsche Mause
(German Fancy Mouse and Mongolian Gerbil Association)
http://www.dmrm.de

SVEMUS (Swedish Mouse Club)
http://www.svemus.org

Glossary

A

agouti - The coloration of wild mice. Each hair is comprised of bands of color, but at a distance the animal appears brown.

albino - An individual characterized by a complete lack of pigment in the skin and hair.

B

bedding - The material used as substrate in the bottom of a mouse's cage. The most typical types are ground corn husks, wood shavings, and shredded newspaper.

C

crepuscular - Animals that are most active at dawn and dusk are said to be crepuscular, but mice are actually nocturnal, showing the greatest level of activity at night.

F

fancy mice - Mice that are bred according to specific programs and standards to achieve a particular look and form and that are considered domesticated and suitable to live as companion animals.

feeder mice - Mice raised in large numbers and often in deplorable conditions that are sold as live feed for snakes, large lizards, and other similar pets.

G

gestation - The duration of a pregnancy from the point of conception to the birth of the young.

gremlin - A gremlin mouse has one normally placed ear, and one that is on the side of the head.

H

herbivore - Herbivores are animals that derive their daily nutrition from consuming plants.

hypoallergenic - Animals or substances that are unlikely to cause an allergic reaction. This is not true of mice.

L

laboratory mice - Mice bred for and kept in laboratories for use in medical and scientific research and experimentation. One of the earliest uses of lab mice was to further the genetic theories of Gregor Mendel.

litter - The term litter can either refer to several offspring born at one time to a single parent or to the substrate used to line the cages of small companion animals like mice.

M

mouse - A small mammal in the family Rodentia characterized by a pointed snout, small rounded ears, and a hairless tail.

N

nest box - A nest box is a cardboard, wooden, or ceramic "house" in which a pet mouse sleeps, hides, and plays.

nesting material - Soft fibers, usually cotton, given to pet mice for the purpose of lining their burrows and sleeping areas.

nocturnal - Animals that are primarily active during the night are said to be nocturnal.

P

pattern - A fancy mouse displaying an even or broken pattern in its coat.

phenols - Phenols are oils that are present in pine and cedar shavings that emit fumes that cause respiratory problems and liver damage in many small animals including mice.

R

rex - Mice with crimped or curled whiskers and fur that develops in waves of varying intensity depending on age.

rodent - A gnawing mammal in the order Rodentia, which includes rats and mice, but also squirrels, hamsters, porcupines, and similar creatures. Rodents comprise the largest of all the orders of mammals.

rosette - Mice that have a whorl of fur on each hip spiraling in the opposite direction of their coat.

S

self - A fancy mouse whose body is all one color.

substrate - Another term for the material used to line the bottom of a small animal's habitat. Also called bedding or litter.

T

tan - A fancy mouse that is a solid color on the upper half of the body with a tan or light colored belly.

W

wild mice - Undomesticated mice that have agouti coats, but appear, at a distance, to be brown. Typically wild mice do not make good pets because their instinct for wariness is too difficult to overcome.

www.ingramcontent.com/pod-product-compliance
Lightning Source LLC
Chambersburg PA
CBHW082111220526

45472CB00009B/2143